Menominee Tribal Enterprises: Sustainable Forestry to Improve Forest Health and Create Jobs

CASE STUDY

PREPARED BY:
CATHERINE M. MATER

A Case Study from "The Business of Sustainable Forestry"
A Project of The Sustainable Forestry Working Group

Contents

I. Introduction 9-1

II. The Business of SFM 9-2
From the Forest 9-2
SFM Forestry Objectives Produce Impressive Results 9-3
SFM Logging Practices Prove a Cut Above 9-7
Logger's "Lottery"—A Winning Solution 9-7
To the Mill 9-8
Constraints to Production Using SFM-Harvested Material 9-9

III. Selling Certified Wood—Not As Easy As It Seems 9-10
The Challenge of Filling Orders 9-10
Veneer Logs 9-10
Lumber 9-11
New Opportunities Require Doing Business Differently 9-11
Market Analysis Leads to Changes 9-12
Recognizing an Advantage 9-12
Conner AGA 9-13
Gibson Guitar 9-13

IV. Certified SFM—Meeting the Bottom Line 9-13
Beyond the Bottom Line 9-13
Good Business Practices a Key to SFM Success 9-15
Evaluating Product Pricing Practices 9-15
Converting Green Lumber Sales to Kiln-Dried Lumber Sales 9-16
Developing Custom Grades for Targeted Species 9-16
Future Business Strategies 9-16
Improved Production Goals 9-16
Increased Marketing and Public Relations Tools 9-17
SFM Prompts Social Pride and Continuing International Education 9-17
Presidential Recognition 9-18
Development of Sustained Development Institute 9-18
Initiation of Sustainable Forestry Demonstration Workshops 9-18
College Curriculum Development 9-18
Development of Student and Youth Programs 9-18

V. Lessons Learned 9-19

Menominee Tribal Enterprises: Sustainable Forestry to Improve Forest Health and Create Jobs

Introduction

The Menominee Tribe has lived in northeast Wisconsin and on Michigan's Upper Peninsula for generations, where ancestral tribal lands once encompassed more than 10 million acres. Following several treaties and land cessions, the Menominee people established a Reservation in 1854 totaling 235,000 acres of predominantly timber land. Since then, the backbone to the economy of the Menominee Nation has been its forests and the industry surrounding the sustainable management of that resource.

The Menominee Tribal Enterprises (MTE) has been an engine of the Menominee economy over the last 140 years and, within the last 25 years, has pioneered the implementation of sustainable forest management (SFM) throughout the Menominee Forest.

Today, the Menominees remain the only Native American tribe to have their forestlands independently certified as being sustainably managed. They are also the only forestlands operation in the United States and Canada that holds dual environmental certification from both the Forest Stewardship Council-approved SmartWood and Scientific Certification Systems (SCS).

The concepts of sustainability in forest ecosystems and surrounding the communities that the Menominee have practiced for so many years include three components of a sustainable forest system:

1. The forest must be sustainable for future generations.
2. The forest must be cared for properly to provide for the many varying needs of people over time.
3. All the pieces of the forest must be maintained for diversity.

Looking closely at what MTE has accomplished in SFM and product development during the last twenty-five years provides unique insight into the economic opportunities and constraints that face other forest products operations considering SFM practices. With a twenty-five-year track record, MTE is one of the few examples in the world where realized forest management performance over time can be compared with intended results to determine whether SFM actually does what it is purported to do:

- Increase the *quality* and *volume* of wood grown in a forest system over time.
- Provide more *consistent* and *stable* annual harvested timber volumes while maintaining or improving forest ecosystems.
- Maintain or improve *a forest ecosystem health* that recognizes the value of multiple uses of a forest.
- Sustain *communities* that surround the forest through job generation and the creation of educational opportunities.
- Increase the *value per unit* of wood products produced from SFM forest resources through documented performance in the marketplace.

MTE's forest management choices may not apply to all forest products concerns. MTE's management and decision-making structure does not appear to be well suited to the management of larger private forestry operations in North America and Europe. It could, however, be applicable to forest businesses owned and/or operated by other tribal or native entities throughout North and South America, and smaller privately-owned forest products concerns worldwide. Equally important, MTE's process of managing tribal forests and the techniques it uses may be well suited for managers of public forestland throughout the world, especially those required to balance the multiple use of forests and deal with the issues of community and public stakeholder trust in the management of the forests.

The Business of SFM

MTE incorporates both forestry and forest products operations in its total business package. However, to understand the business definition and strategic intent of sustainable forestry and sustainable forest products development at MTE, it makes sense to evaluate the forestry operations separately from the sawmill facility. The criteria that provide the foundation for business operations for each are detailed below.

From the Forest

In 1884 about 1.2 billion board feet of standing timber was documented on the Menominee Forest. Since then the amount of billion board feet of standing timber has increased 40%. Moreover, since 1854, even though over 2.25 billion board feet of timber has been harvested off the Menominee Forest, the number of high-quality and large-diameter timber has significantly increased.

The Menominee forest contains a higher diversity of tree species than surrounding forests, which makes using SFM practices more challenging than in other less-diverse areas. The Menominee lands contain approximately eleven of a total of fifteen plant associations found in the entire state of Wisconsin. The major soil types range from dry, nutrient-poor sites that grow poorer-quality tree species (scrub oak, jack pine) to moist nutrient-rich sites that support the growth of very high-value species such as sugar maple and basswood. More than 9,000 distinct timber stands make up the Menominee Forest.

The forest management objective of the Menominee Tribe is to maximize the quantity and quality of sawtimber grown under sustained yield management principles *while maintaining the diversity of native species.* The Tribal leaders have recognized the need for economic survival but only at a speed or intensity that allows for the forest to regenerate itself. Unlike more traditional forest management regimes, this policy promotes a timber harvesting practice that removes timber according to the vigor of the trees rather than their saleable size only. Vigor is defined as the measure of the growth potential of an individual tree. It describes the health and ability of trees to respond to management. Today, Menominee forestlands retain a larger variety and volume of larger-diameter trees on a per acre basis than surrounding forest systems, including federal forest systems next to the tribal lands.

In accordance with their directive to maintain the diversity of native species, the Menominees employ a new Forest Habitat Classification System that provides them with a method to accurately assess the productivity of forest sites and identify the best forest cover for each site, regardless of the tree species or tree quality currently growing on the site.

Traditionally, forest management decisions are based on the current appearance or condition of a forest stand without considering past events such as clearcutting practices. The silvicultural system used by the Menominees recognizes that disturbances on a site and less appropriate harvest practices of the past have dissipated native species and allowed other lower-value scrub species to overtake prime forestland. The system also recognizes that while many species of trees will grow on multiple sites,

MTE Forest Facts

Menominee's Land Includes:

- 235,000 acres of which 94% is covered by productive forestland
- Dominant tree types in the forest are varied and include: northern hardwoods such as maple, red oak, and basswood; hemlock; three types of pine; aspen; and lower quality oak

Source: MTE

Figure 1

Examples of the Most Desirable Forest Cover Types on Menominee Forests

Featured Cover Type	Objective Species	Associate Species
White Pine Mid-Tolerant Hardwoods Red Oak	White Pine Red Oak White Ash Basswood	Red Maple White Birch Quaking Aspen Pin Oak White Oak
Sugar Maple	Sugar Maple	Yellow Birch Hickory White Ash Red Oak Basswood Hard & Soft Elm

The matching of tree species or cover types to a particular habitat type is based on: a) sawtimber growth potential in quality and quantity; b) biological/ecological suitability to the site; and c) competitiveness with other tree species commonly associated with it.

Source: MTE

Figure 2

they achieve their best form and quality on only one or two habitat types. The Forest Habitat Classification System can predict the forest cover type best suited to a site by identifying the existing forest and linking these tree species with a specific habitat. Some species, however, grow better on more than one or two habitat types. White pine is one of those that can thrive on five habitat types.

Using this new system, MTE has identified over 66,000 acres of forestlands that are currently growing below maximum quality and quantity potential. Examples include acreage of aspen, white birch, red maple, and scrub oak growing on sites with habitat types more suitable for pine and quality hardwoods. These acreage once supported high-quality hardwoods and pine but were altered by past management practices that included clearcutting for railroad logging and uncontrolled slash fires caused by poor logging practices. Restoring the 60,000-plus acres currently growing below maximum potential is one of the tribe's primary forestry objectives. The benefits of achieving this objective are listed here.

- The low-value material growing on the site is usually converted into pulpwood that is processed off the reservation. Few on-reservation jobs are supported by this low-quality material.
- The tribal forestlands are converted back to native conditions with larger-diameter, larger-volume, more valuable material constituting the forest landscape.
- Higher-value pines and hardwoods are all processed in the Neopit, Wisconsin, sawmill that is owned and operated by tribal members.
- The higher value species can be better directed toward not only higher profit-per-unit secondary wood product markets, but also markets that prefer certified wood over noncertified wood for product development and may pay a premium for that certified wood source.

Over 6,000 acres are selectively marked for harvest each year. Reentry for cutting in the hardwood stands occurs every twelve to fifteen years. The MTE "order of removal" tree marking rules ensure that a wide diversity of tree species in all size classes and ages are maintained throughout the forest.

SFM Forestry Objectives Produce Impressive Results

With twenty-five years of SFM practices in the forest, the Menominees performance in achieving their intended results ranks high. According to an independent evaluation conducted for FSC-approved certification (SCS and SmartWood) on the Menominee forest in 1994, independent evaluators stated:

> Aesthetically, the Menominee Forest has no equal among managed forests in the Lake States region, although its total productivity measured in the value of the products removed greatly surpasses the adjacent Nicolet National Forest which has more than twice the acreage of commercial forestland.

Because of the unique SFM practices employed by the Tribe over the last twenty-five years, complete forest inventories conducted in 1963, 1970, 1979, and 1989 indicate that the Menominee Forest contained more volume of timber for each successive inventory period, even though over 600 million board feet of timber had been harvested off the forest during that period of time. Equally significant, evidence is clear that the quality of the timber stands and the value/volume of wood growing in the Menominee Forest improved dramatically in several critical areas.

Although the annual allowable cut (AAC) in the Menominee Forest actually decreased from 29 million board feet (mmbf) in 1983 to 27 mmbf in 1995, as noted in Figure 3 the annual harvest remained fairly constant during a thirty-year period between 1962-1995. In addition, SFM techniques used by the Tribe during that time actually increased the *quantity* and *quality* of trees left standing in the forest.

Average Annual Harvest from the Forest (mmbf)

	Sawlogs	Pulp Conversion	Total
1962-69	20.2	4.9	25.1
1970-79	16.1	7.2	23.3
1980-89	13.4	8.4	21.8
1990-95	15.7	6.7	22.4

Source: MTE

Figure 3

Between 1963 and 1988, forestry data documents the conversion of targeted acreage to higher-quality native species matching the appropriate habitat type conditions. The MTE continuous forest inventory (CFI) information shows that the acres covered by lower-quality, lower-value species, such as aspen, decreased, while the acreage covered by the native higher-value species, such as northern hardwoods, increased (see Figure 4).

Cover Type Conversion Trends (1963-1988) on the Menominee Forest (in acres)

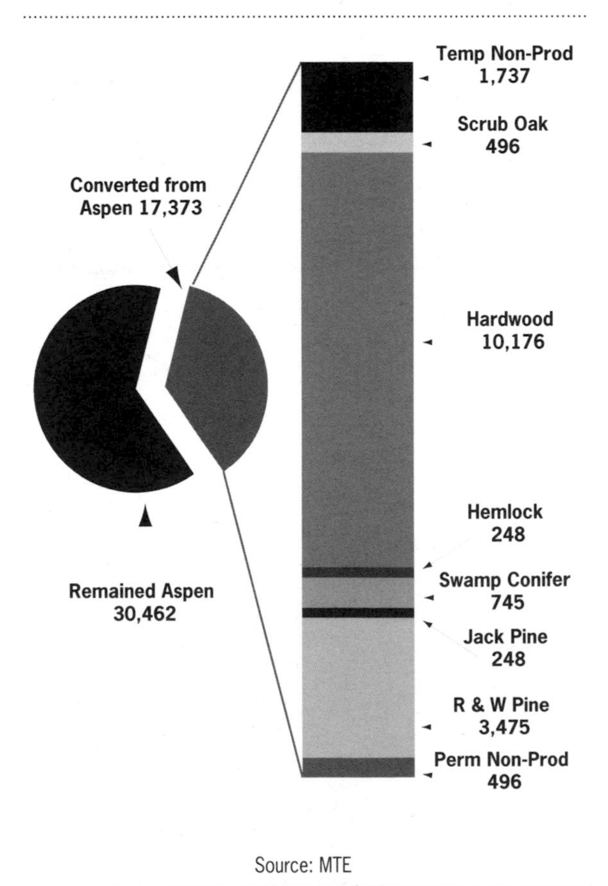

Source: MTE

Figure 4

During the same period, 1963 and 1988, the quality of almost all species represented in the forest system improved. In those same years, the total volume of sawlog (vs. pulpwood) timber resources growing in the Menominee Forest increased by over 200 million board feet, as well.

Through SFM practices, MTE's performance in increasing overall diameter sizes of standing timber within their forest system is exemplary. The quality and grade of sawlogs is often a direct correlation to the diameter size of the tree. Within limits, the larger the diameter of a tree at breast height, the better the grade of lumber material that will processed from that tree with correlating less defect. Measuring by the criteria of diameter at breast height the amount of acreage containing large northern hardwoods sawtimber increased 30% on MTE land between 1963 and 1988 (see Figure 5).

Acres of Standing Timber by Size Class

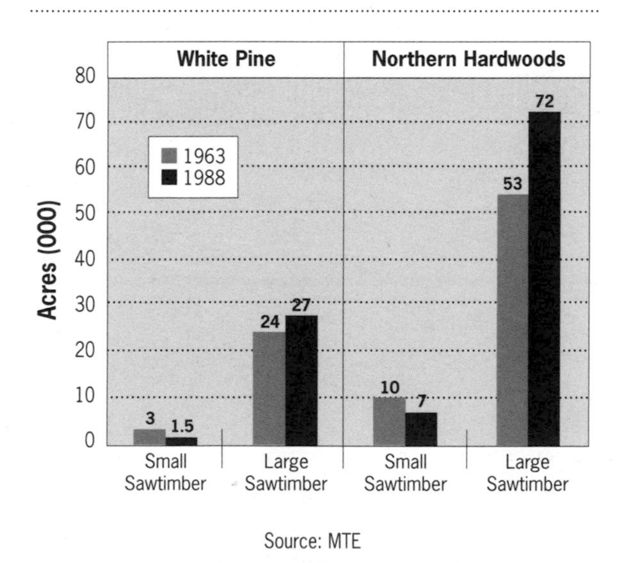

Figure 5

The increase in the diameter size of standing timber has translated into a greater volume of higher-grade wood that is processed into product at the Menominee sawmill. Grade 1 sawlog, for example, is often processed as higher-grade lumber for value-added products such as furniture. Lower-grade material from Grade 3 sawlogs might be processed into pallet stock. Between 1963 and 1988, MTE has increased their Grade 1 sawlog volume from 25% of their total growing stock to over 30% and has decreased their Grade 3 sawlog volume by an almost equal amount.

MTE has been so successful in increasing higher-grade material in its forest that it is currently considering developing three new classifications exceeding standard Grade 1 status. Currently, Grade 1 sawlog applies to at least a 14" diameter log with less than 13% defect. MTE is looking at new Grade 1 classifications that separate up to a 24" diameter.

One of the benefits of SFM practices is the ability for the Menominee Forest to provide more consistent flows of wood volume to the MTE sawmill, although *flow value* can vary from year to year. The sawmill knows in advance what volume it will be processing for the year, and what the grade or quality of material is likely to be. It is then contingent upon the marketing acumen of the sawmill operations to identify the best product distribution channels for both higher-value and lower-value material that can produce the best bottom-line results. This is particularly important with SFM harvests. These harvests avoid high-grading of the forest, or harvesting only the high grade material. Instead, they harvest a balance of lower-grade and higher-grade material. The annual sawlog harvest volume shown in Figure 6 highlights MTE's performance in this area. It shows that between 1970 and 1990 the mill had a more even sawlog harvest volume flow as a result of SFM practices.

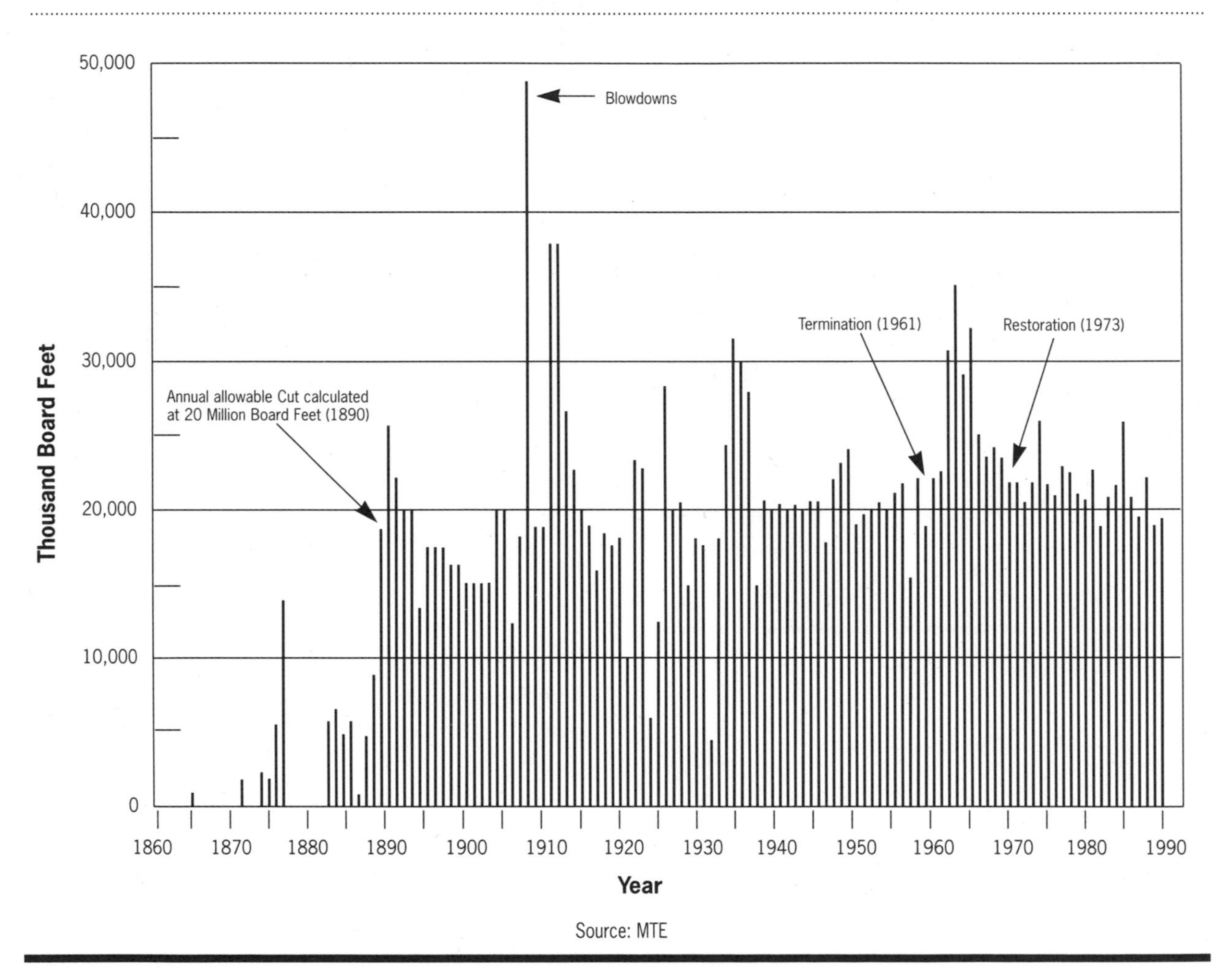

Figure 6

Unlike most other Native American Indian Tribes, and because of its proven track record in sustainably managing its forestlands over the years, the Menominee Nation, as part of its restoration process, was successful in negotiating an agreement with the federal government in 1975 that granted the Tribe the right to manage its forests. Typically, the Bureau of Indian Affairs (BIA) is assigned the duties and responsibilities of managing tribal forests. The tribe is provided a $1.3 million annual subsidy from the BIA to manage its lands. However, that agreement did come with a provision that the Tribe cannot sell or trade their forestland without congressional approval. The land cannot, therefore, be used as a financial asset, which obviously limits MTE's business financing options.

SFM Logging Practices Prove a Cut Above

Performance requirements for loggers contracting to harvest from the Menominee Forest must follow stipulated performance requirements not traditionally seen in logging contracts. The tribe incorporates both financial incentives and penalties for performing or not performing to required standards which ensure sustainable management practices in the field.

Those incentives and penalties include a bonus of $2.50/mbf of logs that are delivered to the MTE for the logger who successfully cuts 100% of the contracted harvest area in accordance with contract specifications. This is a critical incentive as the harvest from the forest includes all grades of material. It is essential to harvest lower-quality material to increase the growing space for higher-quality material in a sustainably-managed forest system. Traditionally, loggers will desire only to harvest and haul larger-diameter, higher-grade material because they get higher prices for it at the mill, or to harvest and haul straight pulpwood because they do it quickly with little regard for damage to the log. Under conventional contracting methods, it is less financially advantageous to the logger to incorporate both.

If a contractor fails to harvest the entire area, barring natural forces or conditions beyond human control, the contractor will forfeit all or a percentage of the performance guarantee based upon the percentage of the remaining contract that is unfulfilled.

Other financial penalties that are designed to ensure sustainable practices in accordance with contract requirements address the skidding and marking of logs, and excessive damage to logs during harvesting as indicated in Figure 7.

Also unique to MTE forestry practices, the logging contracts provide a $5/mbf incentive for the logger to use new thinning techniques in the field that require the logger to extract approximately thirty more "stems" (waste trees between 1"-5" diameter). This allows better growth potential for seedlings.

Examples of MTE Logging Performance Penalties

Action	Penalty Imposed
Unskidded or unmanufactured logs; logs manufactured into pulp	Double the logging rate of Agreement
Cut or girdled unmarked sawlog size tree	$250.00 per tree
Excessive damage to sawlog size tree	$125.00 per tree

Source: MTE

Figure 7

These incentive/penalty stipulations in the logging contracts have helped produce an exemplary record for reduced damage to the forest due to poor logging practices. A report to the Menominee Tribal legislature in 1984 conducted by the Wisconsin Department of Natural Resources found that the Menominee Forests experience 1.9 trees/acre of logging damage compared to 13 trees/acre on national forest systems in the region. This same report concluded a net board foot growth rate per acre per year in the Menominee Forest of 244 board feet compared to 235 board feet of national forest systems in the area.

Logger's "Lottery"—A Winning Solution

MTE also uses a unique contracting procedure for issuing logging contracts. The new "Loggers Lottery" as MTE forestry officials call it, is intended to increase performance on logging contracts and increase the responsibility on individual loggers to complete all their contracts during the year. The lottery process deviates dramatically from traditional logging contract bids, which normally operate on a closed-sealed bid basis, or are assigned by the forest manager.

In the MTE process, however, instead of assigning logging contracts, information on all the sites to be harvested based on certified sustainable management practices is released to all qualified loggers in advance. A date is set for an open bid process. The open bid does not include any price considerations, but concentrates only on the desire of loggers to receive their choice of sites for contract award. If more than one logger desires a particular site, ping-pong balls, one with each logger's name, are placed in a hat and the winning ball is drawn. Once site selection has occurred, negotiations on price begin with the winning contractor.

These negotiations usually end in an agreement. If no price agreement is met, the initial logger is rejected and a new bid process begins. Loggers that successfully gain a contract have a set period of time after the bid process to return selected contracts if they no longer want to handle the site. When this happens, MTE initiates new bids on the returned contract.

MTE gives bidding preference to Menominee Indian contractors. MTE currently has twenty-nine contract loggers who are prequalified to bid on MTE harvest contracts, nineteen of whom are tribal members. Even with the more difficult performance standards attached to logging practices in the Menominee forest, each year there is a growing list of tribal and nontribal contractors who want to contract with the MTE.

The purpose of this new approach is twofold:

- The open no-price bidding process allows contractors to discuss among themselves prior to bid time who should take what contract based on the capability of the contractor to meet performance requirements and avoid penalty payments. More important, the system is intended to ensure that full harvest amounts are completed each year, which is most critical to overall forest health and the viability of the sawmill operations.
- Because contractors select their own sites rather than being assigned sites, they have greater responsibility to perform according to contract stipulations rather than blaming nonperformance on being assigned a "bad site."

The process has worked reasonably well, although there have been problems with some contractors returning contracts at the last minute in hopes of being able to renegotiate a higher price to harvest more difficult and/or lower-timber-quality sites. In recent years, some acreage has not been harvested as a result. In 1995-96, for example, over 1 million board feet of timber was not cut as scheduled. This created a serious shortfall for the mill and negatively affected the health of the forest. In 1997, MTE was changing contract-return times and limitations to correct the problem.

To the Mill

The tribe began operating a small sawmill in the late 1850s. Today, the tribe owns and operates one of the largest sawmill complexes in the region, which employs approximately 160 people in Neopit, Wisconsin. The operation also employs a large team of forest management professionals at its Forestry Center located in Keshena, Wisconsin.

The mill manufactures primary commodity lumber products in both hardwoods and softwoods grown from the Menominee Forest. Standard lumber sizes produced by the mill range in traditional length and width lumber pieces. The mill also sells graded sawlogs and high-grade sawlogs for the manufacture of quality veneers. The approximate percentage splits for resource distribution and product production at MTE are as follows:

- 75%—Logs for lumber production
- 16%—Logs for direct log sales to lumber customers
- 7%—Veneer log sales
- 2%—Log inventory carry-over

The sawmill does not pay stumpage costs to the forestry operations for the logs harvested from their forests. This is a typical operating cost for many private forest products operations that do not own their own forestlands. However, all forestry costs are paid by the mill through product sales.

The sawmill receives no federal subsidies for its operations. Its success is critically dependent on the steady flow of timber from forest to market. The mill only produces from Menominee Forest resources and relies on no outside timber for lumber production and veneer and sawlog sales.

Although new dry kilns were installed in the mill operations in 1996, which provide much needed and valuable increased drying capacity and quality, MTE sells a significant portion of its lumber as green (nondried) to its customer base. Lumber drying is costed at MTE as a service to the client, rather than a high-value dried lumber product offering.

The mill follows product standards and traditional grading rules for both hardwoods and softwoods. Development of custom grades that could convert traditional lower-grade, defect material into higher-value character wood custom grades remains an unexplored opportunity for the mill.

MTE Specie Mix as a Percentage of Production

	1995-1996	1996-1997 (Projected)
Total Production:	10,798,482 (bf)	9,266,940 (bf)
White Pine	26.6%	7.4%
Hemlock	4.0%	14.6%
Hard Maple	14.1%	36.9%
Basswood	6.7%	16.3%
Aspen	6.3%	4.2%

Source: Mater Engineering based on MTE data

Figure 8

Constraints to Production Using SFM-Harvested Material

While SFM practices produce positive, stabilizing results for a forest system, short-term impacts to a sawmill relying on SFM-harvested material prove to be significantly more challenging. Converting forestlands back to higher-grade native species and avoiding high-grading logging practices in the forest are but two SFM mandates that can wreak short-term havoc on a sawmill. The variations in species mixes and in total volumes per species per year that those practices can produce can have dramatic impacts on the business bottom line. Between fiscal years 1995 and 1996, for instance, the MTE sawmill experienced a 17% reduction in high-value hard maple harvest from its forest and a 47% increase in much lower-value aspen harvest, which resulted from an SFM forestland practice of converting acreage to mixed native species. Since the market price for hard maple is about $1,100/mbf more than for aspen, the short-term constraints experienced by the sawmill are dramatic (see Figures 8 and 9).

Differences in the mix of species as indicated in Figures 8 and 9 are especially important considering that white pine prices are historically higher than those for hemlock lumber, and hard maple and basswood are preferred hardwoods for product manufacturing—a fact that is reflected in the consistently higher prices those species command over

MTE Specie Volume Harvest Variations

	% Difference from Previous Year				
	1991-92	1992-93	1993-94	1994-95	1995-96
Hard Maple	Base line yr.	–41%	+5%	+51%	–17%
Aspen	Base line yr.	+49%	+7%	+<1%	+47%

Source: Mater Engineering based on MTE data

Figure 9

other hardwoods such as aspen. In addition, hardwoods, in general, fetch higher prices per unit then softwoods. For MTE, this means that softwood lumber sales for fiscal year 1996-97 should decrease in overall dollar volume, but hardwood lumber sales (depending on the grade being offered) should increase rather significantly.

These species variations from year to year due to SFM practices have substantial impacts on the financial viability of a milling operation and underscore the need to maximize efforts to market and add value to the wood available for harvest in any given year. More clearly put, to afford SFM practices in the forest, it is essential for an operation to practice good business in production, marketing, and sales.

Selling Certified Wood—Not As Easy As It Seems

Although market research conducted by Mater Engineering in the United States during 1996-97 has documented a demand for certified wood that exceeds supply, it appears that MTE finds the selling of certified logs and wood products not as easy as it would seem.

Even though all MTE products come from a certified source, only 4% of hard maple veneer quality sawlogs sold by MTE are actually sold as certified. However, those certified veneer logs command a 10% free-and-clear (above cost plus standard markup) premium. The premium is clearly noted on the customer invoice with the following notice.

> 10% charge for product delivered from certified well-managed forestry operation. SCS does hereby certify that an independent chain of custody has been conducted at MTE and that this facility has been shown to meet certification requirements.

Hard maple and basswood lumber have also been sold as certified but with little documentation showing any premiums attached to those sales. This may be because the accounting procedure for lumber sales does not break out certified sales from noncertified sales, as does that for veneer log sales. However, an MTE sales manager stated that certified lumber sales make up about 5% of MTE's total annual lumber sales. Some of that material fetches premiums of approximately $50/mbf. However, it appears that only the hardwoods can command the premium, which equates to about 4% to 5% for certified hardwood.

The Challenge of Filling Orders

Both the sales managers of veneer and lumber at MTE report an increase in the number of inquiries asking for certified logs and lumber coming from all over the world. For different reasons, however, they face challenges filling those orders.

Veneer Logs

The problem stems *not from a lack of orders but from an abundance of orders and volume that MTE does not have the wood volume to fill.* Unlike high-grading practices that are often employed in standard forestry operations to meet demand, MTE's sustainable practices of pulling set volumes of high-to-low grade material from its forest on an annual basis prevents substantial sales of certified wood that could get 10% premiums. During the research for this case study, for example, MTE received a sales inquiry from a substantial U.S. veneer operation that wanted to purchase 4 million board feet of certified veneer logs annually. Based on SFM practices, the total volume for all veneer logs sold by MTE during 1995-96 was only 540,720 board feet. The projected volume for all veneer logs to be sold in 1996-97 is 814,550 board feet. In essence, the demand for certified veneer logs by one inquiry alone exceeds MTE's capability to supply it annually by almost 400%.

Because of the importance of offering certified veneer log sales to MTE's bottom line (the value per unit of veneer sawlog sales compared to lumber sales is 2:1), veneer sawlog sales are usually negotiated a year in advance, with first buying preference given to certified veneer sawlog buyers.

Lumber

Excluding the exceptions mentioned above, many of the inquiries that MTE receives are for small orders that are not cost effective to supply. Although, as with MTE veneer sawlog sales operations, between six and ten inquiries for certified lumber are received each month, inquires for certified lumber often range below 1,000 board feet per order. Without evaluating more creative product transportation options such as freight load matches and reverse shipping capability at different locations throughout the United States, manufacturers typically prefer full truckload orders to keep per-unit prices competitive for consumers. Standard truckloads volumes per species are 11,000 board feet for hardwoods, and 20,000 board feet for softwoods.

Another problem appears to be MTE's ability to supply consistent volumes of certified wood per grade requirements *on a year-to-year* basis for manufacturers that want to use certification as a marketing strategy for their products. Since MTE is only one of a handful of sources for certified timber in the entire northeastern section of the United States, and with sawlog grades for sustainable harvesting varying from year to year, lack of other certified wood resources in the area that a manufacturer could rely on to ensure consistent quality and quantity of certified resource must be a strong consideration. One advantage, however, of certified forestry is that those harvest evaluations are done several years in advance. This means that volumes and grades of material scheduled for harvesting are also known well in advance. This does allow a level of resource information that MTE can rely on over time for sales and marketing.

New Opportunities Require Doing Business Differently

Due to the visibility MTE has received for its sustainable forestry practices in the Menominee Forest, the Knoll Group, a well-known furniture manufacturer, approached the sawmill in 1991 to consider becoming independently certified. The Knoll Group wanted to purchase and use Menominee Forest hard maple in a commercial furniture line that would be marketed as furniture made from "sustainably-managed maple."

The Knoll Group paid FSC-approved SCS, out of San Francisco, to undertake an independent assessment of the MTE forest management practices. Based on that in-depth assessment, both MTE's forestry operations and the sawmill were certified. MTE became the first forestry and forest products operation to be awarded certification status in the United States. In 1994, the operation underwent another assessment by SCS and a new assessment by SmartWood, another FSC-approved certifier located in Vermont. Based on those assessments, MTE became the only forestry and forest products operation in the United States to receive dual certification. It remains so today. It is also the only Indian tribe in the United States and Canada to hold certification for its forestlands and forest products operations.

While opportunities for selling certified wood do open, as evidenced with the Knoll Group, MTE has discovered that paying for sustainably-managed forests and taking advantage of a certified product offering difference in the marketplace puts added demands on the sawmill and requires management to do business differently. For MTE, this may mean shifting a percentage of its product offering to a new customer base, which is a risky proposition for any operation. MTE has relied most heavily on customers located in its region, and has only recently engaged customers in other parts of North America, Asia, and Europe. Since the demand for certified wood product is greatest in European markets, MTE needs a stronger marketing effort for that part of the world.

Market Analysis Leads to Changes

The mill, which recognizes the need to improve sales and product offerings, mounted a significant mill modernization program in 1995 that has included a complete evaluation of its manufacturing productivity and a new focus on implementing value-added manufacturing in the milling operations in Neopit. The mill also conducted a marketing analysis to improve its performance with existing customers and increase its customer base. A portion of the markets analysis conducted in 1995 was targeted to expanding opportunities for certified MTE wood. The purpose of the survey was to identify ways to:

- Increase the value of certified wood resources to MTE consumers. Up to this point, MTE had not campaigned actively to sell certification of their wood as a unique product offering to customers.
- Identify if an opportunity existed to create custom grades of lumber from wood that is classified as defect material by standard grading rules. This issue was important for evaluating MTE's older-growth hard maple that was scheduled for harvest under certified harvest plans. The old-growth hard maple has a unique brown color configuration ribboned in its sapwood. This "brown stain" is graded as defect material under traditional grading rules based on appearance only (hard maple is often used in sports flooring which requires clear, vertical-grain lumber). MTE wanted to find out if there were opportunities to create a custom grade of lumber from this material.
- Identify if opportunities existed to incorporate added-value in the mill's product manufacturing that could meet customer needs and increase value per unit of timber produced to the mill.

Approximately 20% of the total MTE customer base was directly interviewed for this survey, which produced the following results:

- All of the surveyed MTE customers indicated a desire to continue or increase business relationship with MTE.
- 55% of the surveyed customers indicated an immediate interest in evaluating potential for MTE to provide value-added product to them rather than just commodity lumber. Customers had the most interest in MTE's ability to add either fingerjointing and/or edge-gluing operations to their existing production line. Customers also indicated a strong interest for MTE to increase their kiln-drying capacity for lumber products sold.
- 32% of surveyed customers indicated an immediate interest in evaluating brown stain maple for their high-value products such as flooring, door parts, and panel products.
- 55% indicated an immediate interest in working with MTE to develop visibility for their product manufactured from MTE certified wood, although the majority of customers were unaware they were receiving certified wood from MTE. Customers rated high-value products such as face veneers, flooring, furniture, and specialty products such as cutting boards and tool handles as those having the best product potential for certified wood.
- 23% of the surveyed MTE customers stated an immediate interest in engaging in all three activities with MTE: a) receiving value-added product; b) marketing certified wood products; and c) working with brown stain maple.

Recognizing an Advantage

Although a significant number of MTE's existing lumber and veneer customers were either unaware of or did not use the certification status as a market advantage for their product development, there were some exceptions.

Conner AGA
Located in Amasa, Michigan, this long-time sports flooring manufacturer has been a consistent buyer of MTE's high-grade hard maple. The company had been tracking MTE's certification process and, during 1996, had its own operations assessed by the SmartWood program to receive chain-of-custody certification, which would allow the company to process certified wood from MTE and sell wood flooring as certified. When Conner received SmartWood certification, the company developed a brochure to tell clients and customers about the "new" product offering.

During the research for this case, Conner AGA informed MTE that it had just secured a contract from Walt Disney World, Inc., for hard maple wood flooring for Disney's new facilities in Florida. Conner indicated that the certification status of its product was a key asset in winning the contract, which will use over 165,000 board feet of MTE SmartWood-certified hard maple.

Gibson Guitar
Using MTE's certified maple, Gibson Guitar has released its first line of wood guitars made from SmartWood-certified wood. Two years in the planning, the first Gibson SmartWood Guitar (a Les Paul standard model), made its debut in a widely-publicized October 1996 concert held in New York City to benefit the Rainforest Alliance-SmartWood Program.

Certified SFM—Meeting the Bottom Line

With the exception of 1996, MTE has operated at a profit since 1991 (see Figure 10) even though it has relied primarily on substantially lower-value green lumber sales; has appeared to price some products at below-published price structures; has not yet taken advantage of custom grade development opportunities; and is still in the process of evolving additional value-added production on site. MTE has done so while pioneering SFM practices that have set standards for forest management that are now being reviewed and implemented by other forest products operations throughout the world.

Beyond the Bottom Line

While not meeting traditional industry standards of reaching 20-25% net profit as a percentage of total sales for comparable sawmill operations, MTE's mill

MTE Financials

	1991-92	1992-93	1993-94	1994-95	1995-96
Total Sales ($)	$9,388,258	$10,840,269	$11,528,901	$12,610,480	$11,214,027
% increase/(decrease) in total sales from previous year		+15%	+6%	+9%	(–12%)
Net Profit/(Loss)	$718,942	$776,849	$1,679,780	$1,274,083	($402,507)
Net profit as % of total sales	8%	7%	14.5%	10%	0
Volume bf produced	10,909,368	12,081,244	10,065,446	10, 460,992	10,798,482
Sales price/bf of production	$.86/bf	$.90/bf	$1.15/bf	$1.21/bf	$1.04/bf

Source: MTE

Figure 10

has, except for the 1996 fiscal year,operated at a profit while employing more than double the number of personnel traditionally used in an operation of this annual volume production.

The MTE sawmill, which produces between 10-12 million board feet of lumber annually, employs about 160 people.Typical hardwood milling operations in the private sector for that size of annual production would employ between 80-90 full-time people. Since labor costs typically constitute over 40% of total mill operating costs, this one deviation in operations makes a substantial difference in the financial viability of a milling operation. However, unlike typical private concerns where increasing the bottom line is the prime objective, MTE objectives are intentionally different. MTE strives to achieve, within certified SFM practices, as many consistent, full-time, family-wage jobs for tribal members as possible while maintaining a positive bottom line from year to year.

President Larry Waukau, in the annual President's Report for 1996, stated those goals clearly:

> Our long-term approach to integrated sustainable forest management and wood products manufacturing (a 150-year horizon) precludes MTE from capitalizing upon market peaks and dips. Instead, our approach sets a long-term course which at times requires us to forfeit short-term gain for long-term sustainability.

In addition, the forestry operations incur additional logging costs by providing financial incentives to its contract loggers for harvesting Menominee Forest wood sustainably.These include premiums paid for increased stem thinning in the forest, and premiums paid for harvesting 100% of the contracted volume.

Finally, unlike traditional forest products operations that use intensive high-grade forestry operations, the variation in grade of logs that MTE must harvest from year to year based on sustainable management practices poses additional challenges to achieving more typical profit levels. Over the last

Correlation of MTE Hard Maple Sawlog Grade to Lumber Grade When Processed at the Mill

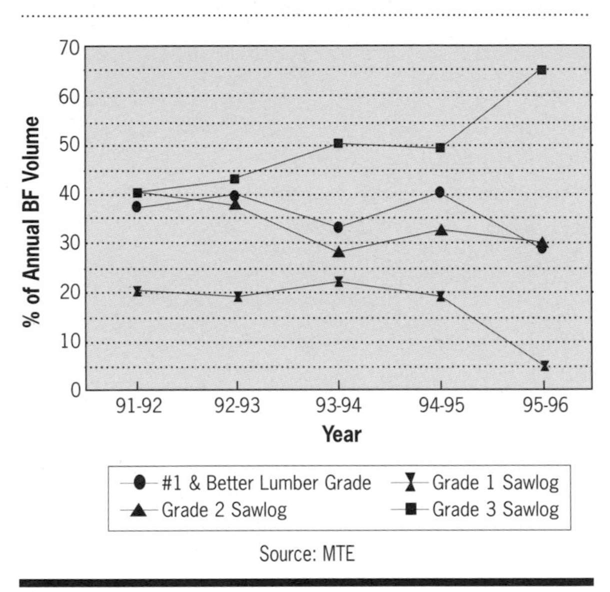

Source: MTE

Figure 11

five years, the sustainable annual allowable cuts from the Menominee Forest have produced dramatic shifts for the mill operations.

Using hard maple volume data from the Menominee Forest as an example, Figure 11 shows that Grade 1 (good quality) logs harvested from the Menominee Forest dropped dramatically between fiscal years 1994-95 and 1995-96. Conversely, Grade 3 (poorer quality) logs increased significantly as a percentage of the overall log volume brought to the mill.The volume of Grade 2 logs remained constant.The impact on the mill is clear: The volume of Grade 1 and better-grade lumber manufactured by MTE, which are used in high-value products, decreased markedly.

While other factors, such as efficiencies in the mill and the accuracy of log grading certainly affect the overall mill production, the short-term variation in harvestable grades necessary to achieve long-term sustainable management practices is a key constraint for traditional forest and mill operations. It clearly illustrates the need to match sustainable forestry practices with other profit-oriented solutions.

The financial loss evidenced by MTE in 1996 was primarily due to weather conditions and unfulfilled logging contracts that left over 1 million board feet of timber standing in the forest that should have been harvested under the certified allowable cut for this last year. However, both forestry and mill operations continued to operate at full employee levels as if 100% of the certified harvest volume had been brought to the mill.

Comparison of MTE Selected Lumber Prices to Standard Published Lumber Prices over Time (prices are for non-dried 1" thick lumber)

	July 1994-June 1995			July 1995-June1996		
	Published Avg. ($/mbf)	MTE Avg. ($/mbf)	% Difference	Published Avg. ($/mbf)	MTE Avg. ($/mbf)	% Difference
Hard Maple:						
FAS	1132	1063.5	(–6.5%)	1141	1027.2	(–11%)
#1C	761	715.3	(–6.4%)	756	683.6	(–10.6%)
Red Oak:						
FAS	1354	1276.6	(–6%)	1252	1304.76	+ 4%
#1C	884	881.6	(–.3%)	818	943.78	+15%
Basswood:						
FAS	769	801.6	+4.1%	749	818	+ 9%
#1C	383	412.1	+7.6%	365	409	+12%

Source: Mater Engineering based on published lumber prices

Figure 12

Good Business Practices a Key to SFM Success

It is clear that to financially sustain a certified operation, there can be no substitute for matching SFM practices with other profit-oriented business solutions such as increasing mill efficiencies, adding value to commodity products, developing custom grade material, and marketing products made from certified wood better.

As they are in most contemporary sawmill operations, efforts to improve production and increase profit margins at MTE are ongoing. MTE has a number of potential opportunities that it could consider to increase sales and improve production.

Evaluating Product Pricing Practices

Product prices on a per-unit basis can vary due to many factors. Long-term relationships with customers, distance of the mill from customer location, and fluctuations in current market values are but a few significant factors that can affect pricing decisions. Even so, keeping track of averaged published market prices can provide some interesting insights into possible missed opportunities. For this case study, published hardwood prices in Hardwood Review during fiscal years 1994-95 and 1995-96 for high-grade lumber were analyzed and compared to MTE pricing data for the same period of time. While MTE appears to be capturing above published prices for certain species such as basswood and red oak, they appear to have also consistently priced below published prices for higher grade hard maple (see Figure 12).

This action is noteworthy considering that hard maple has represented a significant portion of MTE's annual hardwood lumber sales. For 1996-97, hard maple is projected to make up almost 40% of MTE's total lumber sales.

Converting Green Lumber Sales to Kiln-Dried Lumber Sales

Adding value to raw resource is a critical area most mills are exploring today along with MTE. Currently, only about 25% of the lumber MTE sells is sold as dried in its own kilns, the rest is sold as green. However, MTE charges for that lumber drying as a service to the customer at an average rate of between $50-100/mbf of lumber dried. This value is then added on to a green lumber price. The missed opportunity to increase gross revenue to the mill is best seen by comparing published green hardwood lumber prices against published prices paid for kiln-dried lumber as a product offering (vs. a service offering). For the hardwood species such as hard maple, red oak, and basswood that MTE sells, the average percentage increase between green and dried lumber indicates a missed gross profit opportunity of approximately $300/mbf; a 40% average increase in gross revenue per thousand board foot of hardwood lumber sales (see Figure 13). MTE's projected 1996-97 value of $100/mbf added for lumber drying as a service appears to represent only a fraction of the actual income that might be generated by offering dried lumber as a product.

Non-Dried (Green) vs. Dried Lumber Comparison ($/mbf; 1" thick lumber)

	Green (FAS; #1)	Kiln-Dried (Sel & Btr; #1)	% Difference
Hard Maple:			
Sel & Btr/FAS	$1,126	$1,454	+29%
#1C	744	1,079	+45%
Basswood:			
Sel & Btr/FAS	756	1,0571	+40%
#1C	380	607	+59%
Red Oak:			
Sel & Btr/FAS	1,340	1,693	+26%
#1C	869	1,161	+34%

Source: Mater Engineering

Figure 13

Developing Custom Grades for Targeted Species

Although as yet unexplored by MTE, as it is with many other traditional wood product producers, developing custom lumber grades can present significant opportunities for a sustainably-managed forest products operation that must rely on varying grades of material from year to year. MTE's 1995 customer survey indicated interest in this area, especially for hard maple with brown ribboned character in the wood. Under normal conditions, the configured wood would be either considered waste or converted into economy grade material often used in pallet production. Pallet stock usually sells between $125/mbf to $200/mbf. Converting the wood into a custom grade to be used in the flooring markets, for example, could increase prices to over $500/mbf.

Future Business Strategies

Since its increased visibility as a certified sustainably-managed forest products operation, MTE has initiated additional marketing and sales activities that are intended to help maintain its SFM practices in the forests and increase certified wood products sales. Those activities are noted below.

1) Improved Production Goals

MTE has identified the following mandate to complete in fiscal year 1996-97.

- Improve lumber inventory control systems to increase production efficiency and provide better information for consistent product delivery to MTE customers.

- Complete energy audits and related development improvements in energy system designs and development. These improvements will reduce overall production costs and provide additional revenue to sustain a certified wood products operation.

- Continue the study and planning for development of a log sorting yard, whole log chipping, and boltwood processing facilities. These activities should greatly increase MTE's ability to make more with less and help to convert traditional wood waste to wood profits.

2) Increased Marketing and Public Relations Tools

MTE has developed a number of devices that highlight its certified wood product offerings and SFM practices in the Menominee Forest.

- *Brochure visibility.* MTE's facility brochure highlights its commitment to sustainable forest products both on its cover and in the text of the brochure which states:

  > Another added feature for MTE is the recognition by Scientific Certification Systems, more commonly known as "Green Cross Certification," for incorporating sustainable management practices into the overall management of their forest and the contribution protecting habitat specie biodiversity and timber resource base of the forest in which they are harvesting.

- *Video visibility.* Within the last year, the demand for information about MTE's SFM practices has prompted the development of two videos that are now distributed to MTE customers, interested parties, and the public worldwide. *Listening to the Forest* is a fifteen-minute documentary on the management of the Menominee Forest and the way that management has improved the health of the forest and the economy of the Menominee Nation. A more technical video, *Sustainable Management of White Pine Stands,* addresses interests specific to forest managers and concerns about the sustainable management of white pine stands in the Great Lake States area.
- *Establishment of public demonstration sites on sustainable forestry and certified wood products from MTE.* Three sites within the Menominee Forest have been designated as sustainable forestry demonstration sites for the public. The sites will include signage, designated guided tours, safety requirements, and equipment and materials used in sustainable forestry operations. At the main sawmill in Neopit, a certified wood products showcase is to be established, starting with the MTE building itself, illustrating the certified wood products used in the construction of the MTE administration building.
- *Establishment of the MTE World Wide Web site.* Initiated in 1996, MTE has its own Web site to expand Menominee's sustainable forestry and certified wood products visibility worldwide (http://www.menominee.com).
- *Bumper stickers for cars.* The tribe has bumper stickers to distribute to the public that advertise MTE as offering hardwood and softwood. They read *Native American Sustainable Resources.*
- *Annual marketing workshops for MTE customers.* During 1995, MTE initiated its first annual marketing workshop for MTE customers. The annual workshop provides current marketing information about the specific products manufactured by MTE's customers and incorporates a full section on growing markets and product distribution systems for moving certified wood products in international markets.

SFM Prompts Social Pride and Continuing International Education

The changes that have occurred in the Menominee Nation through the implementation of SFM in the Menominee Forest have provided significant social and educational benefits to the People of the Menominee Nation. The visibility of Menominee's work in SFM has strengthened social pride within the Menominee People and increased opportunities for new programs that generate added income to the tribe. Some of these changes and benefits are described here.

Presidential Recognition

In 1995 MTE was one of a handful of U.S. organizations to receive the first-ever United States President's Award for Sustainable Forestry Development. The event, which was one of great pride to the Menominee Nation, honored MTE for its pioneering achievements in SFM practices.

Development of a Sustained Development Institute

Prompted by the success of SFM practices employed in the Menominee Forest, in 1993 MTE and the College of the Menominee Nation jointly established the Sustained Development Institute (SDI) for the Menominee Nation to promote the Menominee Forest and its management, and to educate the public and especially the Menominee children about the resource. The Institute was also chartered to further develop and describe the principles of sustainable development, apply them to different environmental situations, and then to educate the public about the principles and how they are applied. The institute's specific tasks include:

- Providing tools useful for those who want to study and work toward sustainable communities.
- Developing on-reservation demonstration areas to teach visitors about the major elements of Menominee sustainable development.
- Developing a curriculum, working in concert with the U.S. Department of Agriculture, that teaches the elements of Menominee sustainable development and relates those to issues, efforts, and movements in other parts of the world.
- Developing an Internet-based program in concert with the Land Tenure Center of the University of Wisconsin that can lead to a four-year degree in sustainable development.
- Developing short seminars on Menominee sustainable development for visitors, students, professional foresters, environmentalists, and others.

Initiation of Sustainable Forestry Demonstration Workshops

In October 1996, MTE gave its first sustainable forestry demonstration workshop for public and private forestland managers. The conference and field demonstrations, which concentrated on timber harvesting systems for SFM, had over 120 participants from 10 U.S. states and other countries. About 20% of the participants came from the provinces of Canada, with private forestland owners from as far away as South America also attending.

College Curriculum Development

College curriculums are now offered at College of the Menominee Nation detailing cost/benefit economic analyses of sustainable forestry issues. The designed coursework is not only offered to Menominee Tribal students, but is also extended to Department of Natural Resources state forestry staff and U.S. Forest Service personnel. The course is also offered through the Internet with "virtual students" registered from as far away as Sweden. Case studies used in the training and analyses are actual Menominee Forest management cases.

Development of Student and Youth Programs

The MTE has also initiated new programs aimed at educating tribal youth about job opportunities and the values inherent in SFM practices, which include:

- Student-to-Work Programs that target high school students interested in working within the MTE Forestry Division to give them experience and high school credits for hands-on experience in Menominee SFM practices.
- Youth learning programs that allow younger tribal students to spend time in the field with MTE foresters learning about SFM practices. The students are then required to transfer what they have learned to global students through the Menominee Web site (http://www.menominee.com/camp).

Lessons Learned

The experience of the Menominee Nation illustrates both the opportunities and constraints that can arise in the practice of SFM.

The evidence clearly indicates that SFM practices can work—in the forest, the community, and for the bottom line. Since 1854 The Menominee People have lived on their Reservation and intensively managed the land—yielding billions of board feet of timber for production. The wise use of the resource has created hundreds of sustained, full-time, family-wage jobs for tribal members for almost 100 years. But equally important, SFM has increased *quantity* and *quality* of remaining standing timber in the Menominee Forest.

The experience of MTE, however, also suggests that SFM practices come with a price. Short-term gains may need to be sacrificed for long-term sustainability goals. The short-term challenges of producing a profit underscore the importance of conducting good business practice *in tandem with* SFM practices. For MTE, this may mean:

- Reevaluating product pricing strategies.
- Increasing product diversity through increased product offerings (such as dry-kilned lumber).
- Developing custom grades that better market character wood.
- Increasing value-added product offerings that convert wood waste into wood profits.
- Implementing a log sort yard to take advantage of multiple log buyers' needs

Using SFM practices in the forest does make for good business in the mill. While critics may point to the business benefits of being an Indian Tribe (government subsidy; no payment of taxes, etc.) as key components to financial success, it is clear that the Menominees have made specific choices in the way they run their business. This includes *double* the typical level of employees needed for similar-sized forestry and wood processing operations. When one considers that labor wages for a typical wood processing operation the size of MTE account for about 30% of the total operating costs—with taxes accounting for about 8%—the impact of MTE's increased employment goals on its financial performance is clear.

It is unclear whether any premiums attached to the sale of Menominee's certified wood and wood products are, or will be, a growing or even consistent trend to the company or others marketing certified wood products. However, it is evident that in certain product offerings, the current annual demand for certified products exceeds the supply. For this reason, expanding the amount of certified wood available is vital to opening up and gaining access to markets for certified products.

Copies of individual cases studies, or a bound set of all the case studies listed below, are available for purchase from the distributor, Island Press, which in 1998 plans to publish a book-length study based on this material entitled *The Business of Sustainable Forestry.*

For purchasing information contact:

Island Press
Phone 800.828.1302
Fax 707.983.6414

The working group has an Internet web-site at http://www.sustainforests.org

The Cases

Overall Market Analyses:

OVERVIEW OF SUSTAINABLE FORESTRY
A conceptual and illustrative framework for sustainable forestry.

SUSTAINABLE FORESTRY WITHIN AN INDUSTRY CONTEXT
Defines the relationship between sustainable forestry and the entire forestry industry.

MARKETING PRODUCTS FROM SUSTAINABLY MANAGED FORESTS: AN EMERGING OPPORTUNITY
The current demand for sustainable forest products and the likely demand over the next two to five years.

A REVIEW OF EMERGING TECHNOLOGIES
New technologies which influence investment decisions in sustainable forest management.

Business Case Studies on Companies or Landowners:

ARACRUZ CELULOSE S.A. AND RIOCELL S.A., BRAZIL
COLLINS PINE COMPANY, U.S.
COLONIAL CRAFT, U.S.
J SAINSBURY PLC AND THE HOME DEPOT, U.K./U.S.
MENOMINEE TRIBAL ENTERPRISES, U.S.
PARSONS PINE PRODUCTS, U.S.
PORTICO S.A., COSTA RICA
PRECIOUS WOODS, LTD., BRAZIL
STORA, SWEDEN
VERNON FORESTRY, B.C., CANADA
WEYERHAEUSER COMPANY, U.S.

Case Studies on Small Private Landowners:

(7 representative U.S. properties)

BRENT PROPERTY
CARY PROPERTY
FREDRICK PROPERTY
FREEMAN PROPERTY
LYONS PROPERTY
TRAPPIST ABBEY
VAN NATTA TREE FARM

The Business of Sustainable Forestry

Project Director:

Michael B. Jenkins
John D. and Catherine T. MacArthur Foundation
Program on Global Security and Sustainability

Project Coordination and Support:

Greg Lanier
John D. and Catherine T. MacArthur Foundation

Project Contributors:

Matthew Arnold
World Resources Institute,
Management Institute for Environment and Business

Tasso Rezende de Azevedo
Instituto de Manejo e Certificação Florestal e Agrícola

John Begley
Weyerhaeuser Company

Bruce Cabarle
World Resources Institute,
Management Institute for Environment and Business

David S. Cassells
The World Bank, Environment Department

Rachel Crossley
Environmental Advantage, Inc.

Robert Day
World Resources Institute,
Management Institute for Environment and Business

Betty J. Diener
University of Massachusetts, Boston

Richard A. Fletcher
Oregon State University Extension Service

Jamison Ervin
Forest Stewardship Council

Eric Hansen
Oregon State University,
Department of Forest Products

Stuart Hart
University of Michigan,
Corporate Environmental Management Program

Tony Lent
Environmental Advantage, Inc.

Stephen B. Jones
Auburn University

Isak Kruglianskas
Universidade de São Paulo,
Faculdade de Economia, Administação e Contabilidade

Keville Larson
Larson and McGowin, Inc.

Stephen Lawton
Oregon State University, College of Business

Catherine M. Mater
Mater Engineering, Ltd.

Scott M. Mater
Mater Engineering, Ltd.

James McAlexander
Oregon State University, College of Business

Bill McCalpin
John D. and Catherine T. MacArthur Foundation

Mark Miller
Two Trees Forestry

Mark Milstein
University of Michigan,
Corporate Environmental Management Program

Larry A. Nielsen
Pennsylvania State University,
School of Forest Resources

Diana Propper de Callejon
Environmental Advantage, Inc.

John Punches
Oregon State University,
Department of Forest Products

Richard Recker
Oregon State University,
Sustainable Forestry Partnership

A. Scott Reed
Oregon State University, College of Forestry

Jeff Romm
University of California at Berkeley,
College of Natural Resources

Nigel Sizer
World Resource Institute, Biological Resources and
Institutions/Forestry Frontiers Initiative

Michael Skelly
Environmental Advantage, Inc.

Thomas Vandervoort
Vandervoort Public Affairs & Communications

Court Washburn
Hancock Timber Resource Group

Michael P. Washburn
Pennsylvania State University,
School of Forest Resources

Mark Webb
Mark Webb & Co.

Charles A. Webster
Environmental Advantage, Inc.

Peter Zollinger
FUNDES/AVINA Group

Communications and Support:

Ted Hearne
Ted Hearne Associates

Erika Fishman
Ted Hearne Associates

Tess Hartnack
Hartnack Design
Case Study Design

Ray Boyer
John D. and Catherine T. MacArthur Foundation

Emily T. Smith
Environmental Journalist
Case Studies Editor

"The Business of Sustainable Forestry"
A Project of The Sustainable Forestry Working Group

Menominee Tribal Enterprises: Sustainable Forestry to Improve Forest Health and Create Jobs

CASE STUDY

PREPARED BY:
CATHERINE M. MATER

A Case Study from "The Business of Sustainable Forestry"
A Project of The Sustainable Forestry Working Group

The Sustainable Forestry Working Group

Individuals from the following institutions participated in the preparation of this report.

Environmental Advantage, Inc.

Forest Stewardship Council

The John D. and Catherine T. MacArthur Foundation

Management Institute for Environment and Business

Mater Engineering, Ltd.

Oregon State University
Colleges of Business and Forestry

Pennsylvania State University
School of Forest Resources

University of California at Berkeley
College of Natural Resources

University of Michigan
Corporate Environmental Management Program

Weyerhaeuser Company

The World Bank
Environment Department

World Resources Institute

CCC 1-55963-624-6/98/page 9-1 through page 9-19